Alam Zeb

Problems and Prospects of Redd in Pakistan and in Community based Forest Management System

GRIN Verlag

RESEARCH ARTICLE ON REDD ,

"PROBLEMS AND PROSPECTS OF REDD IN PAKISTAN AND IN COMMUNITY BASED FOREST MANAGEMNT SYSTEM"

Mr. Alam Zeb Qazi,(Shaheed Benazir Bhutto University Sheringal Dir Upper kpk Pakistan)

Abstract

Deforestation and forest degradation are the second leading causes of global warming. Interes t in the potential of avoided deforestation has been growing and solutions to the problems are currently being worked to through the development of proposed mechanism under the Kyoto Protocol (KP) called Reducing Emission Thorough Deforestation and Degradation (REDD). This Research-paper argues that unless a significant change

is made in the existing proposal of REDD till Copenhagen in December 2010, prospect of be nefiting PAKISTAN forest in general and community forestry in particular looks not so encouraging. It is mainly due to factors like poor documentation of deforestation and degradat ion (uncertainties of baselines), small scale of forests to attract global buyers and PAKISTAN low influence in the international REDD policy process, among others. This Paper tries to an alyse that, despite achievements in the forestry capital formation, several contentious issues r elated to the equitable sharing of income among different stakeholders have emerged, and ma y even threaten the sustainable management of this capital if issues are not duly addressed. M uch of the debate on forest

based mitigation measures are dominated by technical issues and socio-

ecological systems such as livelihoods and biodiversity tend to take a back seat. Finally, the a uthor has forwarded the responsibility of Gigantic Institute of Forestry like Shaheed Benazir Bhutto University Sheringal and PFI (which are the top most ones)

in carbon inventory process.

INTRODUCTON

REDD: WHY IT CAME? AND WHAT DOES IT MEANS?

Deforestation and forest degradation are the second leading causes of global warming. Tropic al forest clearing accounts for roughly 20% of the anthropogenic carbon emissions and destro ys significant carbon sinks globally (IPCC, 2007). The global response to climate change is c oordinated through the United Nations Framework Convention on Climate Change (UNFCC C) and since early 2005, under the Kyoto Protocol.

Efforts to reduce atmospheric Greenhouse Gas (GHG) concentrations (and thereby mitigate in creases in global temperatures) are being made through schemes that aim to reduce the use of fossil fuels, increase energy efficiency and sequester carbon dioxide in biological matter. In d eveloping countries (especially

PAKISTAN) these schemes are managed in two ways: first, regulated or certified projects wh ich come under the Clean Development Mechanism (CDM) of the Kyoto

Protocol and which are regulated according to international standards; and second, 'voluntary' projects which operate outside of the KP and have no overall governing body, although volun tary standards may be compiled to (Staddon, 2009). These certified and voluntary projects inv olve the trading of Certified Emission Reduction (CER) or Voluntary Emission Reduction (V ER) credits respectively. One CER or VER credit is typically equivalent with the capture of o ne ton of carbon dioxide, and companies or individuals buy credits to comply with legal com mitments for the purpose of Corporate Social Responsibility for philanthropic reasons.

Almost all CDM projects involve energy efficiency or energy reduction, with only one out of the 400 registered projects relating to carbon sequestration through forestry, falling under the 'Land-Use, Land-

Use Change and Forestry' (LULUCF) category. The lack of forestry projects is considered to be due to the high transaction costs of the CDM process and restrictions placed

on forestry projects by the CDM (Peskett, Brown, & Luttrell, Making Voluntary Carbon Mar kets Work Better for the Poor: The Case of Forestry Offsets., 2006a). In contrast, the more fle xible voluntary market is dominated by forestry related projects located throughout the globe (Peskett, *et. al.* 2006a).

Under the CDM's LULUCF Programme, only Afforestation and Reforestation (AR) projects are currently recognized, and many projects in the voluntary market involve the planting of tr ees, meaning that there is a little scope for projects that work through 'avoided deforestation'. Reasons given for the exclusion of the avoided deforestation include the difficulties in ensurin g 'additionality' (i.e. the project provides emissions reductions which are additional to what w ould have occurred in its absence), problems in establishing 'baselines' (levels from which to estimate emission reductions due to the project), problems in preventing 'leakage' (changes i n emissions due to project activities but which occur outside of the project area, e.g. shifting d eforestation to another area), the problem of ensuring that reductions are permanent, the fact t hat the large potential scale of avoided deforestation emission reductions could flood the mark et, and finally, that it may reduce incentives for developed countries to 'de-
carbonize' and reduce emissions through energy efficiency or energy reduction (Peskett *et. al.* 2006b, Richards and Jenkins 2007, Karky and Banskota 2007).

Interest in the potential of avoided deforestation has been growing, however, and solutions to the problems listed above are currently being worked out through the development of propose d mechanism under he Kyoto
Protocol called REDD. At the 13th Conference of Parties (CoP 13) of the UNFCCC, held in Bali in 2007, it was agreed in principle to implement a policy called REDD in developing cou ntries upon completion of the first commitment period of Kyoto Protocol, i.e. 2008-
2013. Under the proposed REDD Policy; there is a strong move to reduce CO2
emissions from terrestrial ecosystems by reducing deforestation rates in the tropics (cited by Karky and Banskota, Gullison *et. al.* 2007). Additionally, at the UNFCCC meeting held in Ac cra from August 21-
27, 2008, REDD gained a broader support for an agreement to reward actions that enhance for est cover and is additional to avoided deforestation and forest degradation (popularly called R EDD+). Once it comes into operation, this will add value to the existing forest capital and the stream of income that flows from it through reduced deforestation, reduced degradation and f orest enhancement, and, hence, should theoretically benefit all primary stakeholders. However , to make claims to this income, it will be essential to demonstrate that there has indeed been a net sequestration of carbon. Other social, economic and technical issues will also have to be c

omplied with. As the policy is still being discussed and negotiated, uncertainties associated wi th REDD implementation will continue to trigger more questions and debates before they are decided to the CoP 15 meeting in December 2009.

MATERIAL & METHODS

The study has carries out in Pakistan and in community based forest Management system Pakistan is comprises of four provinces including FATA & Gilgitbaltistan with northern areas,As the divisional system was abolished in 14 August 2001, there are 117 districts including capital district Islamabad, in which 35 districts in Punjab,24 in Khyber PukhtoonKhwa, 27 in sindh, 30 in Balochistan ,3 districts in federal capital, 7 districts in Northern areas and 8 districts in AJK. In Pakistan there is 4.8% forest cover (inlatest books 4 % by some authors) in which Punjab contributes 0.51 million hectares, Khyber Pukhtonkhwa contributes 1.33 million hectares, Sindh contributes 0.84 Million hectares & Balochistan contributes 1.36 million hectares. (As shown in below Map indicating forests & table indicates their respective cover)

In Khyber PukhtoonKhwa there is no desert, Pakistan 13 % of mountain range is covered by glaciers , it covers 13680 square.km(Siachen glacier 72.5 km, Hasper glacier 61 km). Length of coastline of Pakistan is 1046 km,58 % area of Pakistan is covered by Mountains & plateaus. Rajoka ,A.,Raza,W,H.,Numan,M,A,.(2012)

Pakistan raised major concerns about the protection and growth of forests and active local part icipation in forest management. This can be attributed to the efforts of thousands of CFUGs. Many issues and challenges triggered by the stocks thus rejuvenated have given rise to a num ber of subset of issues, including ownership, rewards and distribution. This often culminated s erious conflicts. This is primarily the second generation issue in Pakistan's community forestr y and needs to be addressed with due priority (Dahal & Banskota, 2009).

To date some of the 14,439 plus CFUGs are managing over a million hectares of forests and a re generating income in most places DoF 2009).

CURRENT % USES OF WOOD IN PAKISTAN

MAJOR USES OF WOOD	PERCENTAGE
Fuel wood	80%
Building construction , Furniture	15%
Railway sleepers, packing, industry(sports, goods, chipboards, match box etc.	5%
Total	100%

(Malik,A,M., & Ahmad,M,R,. (2010)

TOTAL FOREST COVER OF PAKISTAN

S/NO	PROVINCE	TOTAL AREA. Sq.km	FOREST AREA. Million hectares
1	Punjab	205344	0.52
2	Khyber PukhtoonKhwa	74521	1.33
3	Sindh	140914	0.84
4	Balochistan	347045	1.36

(Malik,A,M., & Ahmad,M,R,. (2010)

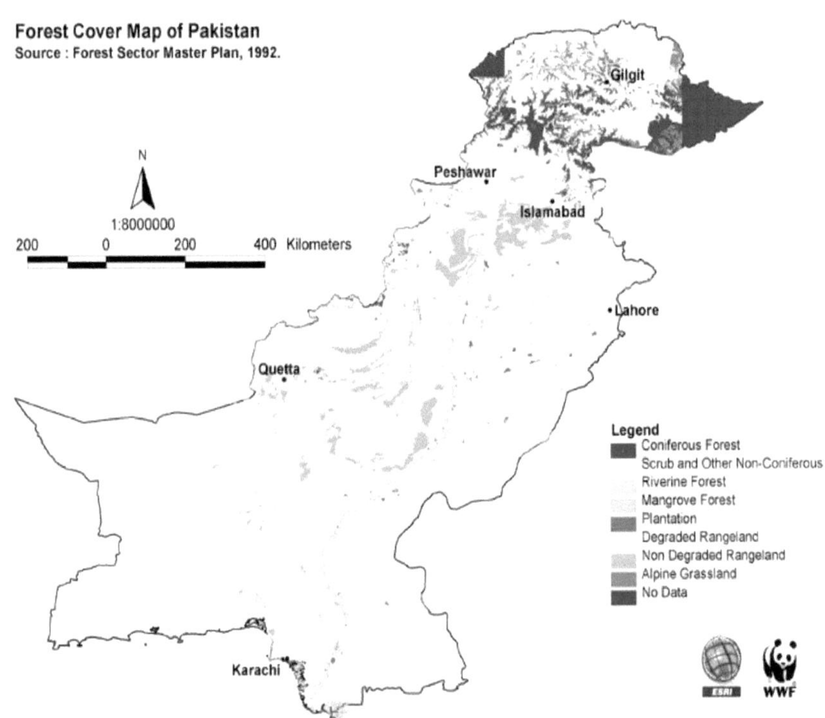

DISCUSSIONS

REVIEW OF NATIONAL EXPERIENCES

Despite Pakistan's very low GHG emission, it's MINISTRY OF ENVIRONMENT (Forest and Soil Conservation) has taken progressive step towards climate change mitigation through market mechanisms offered by REDD. It has ventured to embark o n the World Bank's REDD Programme so as to be ready for forest carbon transaction when th e tenure of the Kyoto Protocol ends in 2012-13.

The World Bank is moving ahead to address methodological and institutional issues related to REDD. The Bank established the new Global Forest Alliance (GFA) in 2007, in partnership with agencies like Nature Conservancy, Conservation International and World Wildlife Fund for creation funds to intervene in the forestry sector. The Bank has been able to establish the F orest Carbon Partnership Facility (FCPF) with a financial plan of US$165 million to jump-start REDD policy in developing countries like Pakistan Nepal's, etc (World Bank 2008). Preparation and submission of the Readiness Plan Idea Note (RPIN) is the first step in this dir ection. The RPIN process has been characterized by:

Exploration of potentials, problems and recommendations associated with the future national REDD framework.

1. Capacity development package to fulfill the operationalization aspects of the above fra mework.

2. Exploration of the impending governance issues related to the operationalization aspec ts.

3. Setting informal multistakeholder processes in motion by developing RPIN through a consultative process.

4. Projecting success in participatory forest management for leveraging on a pro community forestry REDD concept. (Adapted from Rajan Kotru, 2009)

Pakistan participated in the World Bank's call for submission of a RPIN. Based on its experie nce with Community Forest Management, the RPIN document submitted by Pakistan was sel ected by the World countries. Once the RPIN is formulated into a full Readiness Plan (RPlan) , Pakistan will be able to implement a prototype of REDD and will be able to gain experience and build capacity to operationalize REDD by taking onboard community forest users in and experimental way under FCPF , Ministry of environment & other conservative organizations. Though the FCPF is outside of the UNFCCC, the experiences from FCPF may be valuable for the development of REDD Policy under UNFCCC aimed at the post 2012 period.

It may be possible for Pakistan

to be able to lobby for conducive REDD policies based on experiences from RPIN. In January 2009, nine members multistakeholder consultative committee under the coordination of MFS C Secretary has been formed to precede the task of readiness plan with FCFP funds (Dahal and Banskota, 2009).

RESULTS

OPPORTUNITY AND CHALLENGES IN COMMUNITY FORESTRY

In 1979s, the declining forest capital in the hills of Pakistan raised major concerns about the protection and growth of forests and active local participation in forest management. This may be referred to as the first generation issue, which has to do with increasing the capital stock of forests. Since the 1990s, forest capital formation has been fairly successful in the hills of Pakistan. This can be attributed to the efforts of thousands of

CFUGs, which manage their forests as per agreements reached between those and the district forest offices concerned.

Pakistan community forestry is the key element behind the projection of positive image of Pakistan's

overall forest status. While community forestry has addressed the first generation by restoring forestry capital and by devising an institutional mechanism to manage this capital, the story is far from over. Many issues and challenges triggered by the stocks thus rejuvenated have given rise to a number of subset of issues, including ownership, rewards and distribution. This often culminated serious conflicts. This is primarily the second generation issue in Pakistan's community forestry and needs to be addressed with due priority (Dahal & Banskota, 2009).

To date some of the 14,439 plus CFUGs are managing over a million hectares of forests and a re generating income in most places DoF 2009). In many places this capital has eased proble ms of firewood and fodder supplies in mountain areas. Capital formation has created opportun ity for making income. Despite achievements in the forestry capital formation, several arguabl e issues related to the equitable sharing of income among different stakeholders have emerged , and may even threaten the sustainable management of this capital if issues are not duly addre ssed. Claim over property rights on the part of the government gave rise to conflict. For exam ple, the Government of Pakistan (GoP) attempted to exercise its property rights over this capit al by claiming 40% of the cash revenues generated through commercial sale of forest products . Though the government's decision was overturned by the Supreme Court of Pakistan, citing inadequate legal ground over the claim, such conflict may merely be the start.

There is now a new dimension added to the use of forests. The recognition to compensate for carbon sequestration function of forest has added to the potential value of community forest resources..
Once an international agreement is reached on REDD, community forests may be the potentia l source of extra benefits brought about by carbon conserved in their forests. This obviously w ould add monetary value to the existing community forests. After the REDD mechanism is fra med into an international protocol, the debates would shift to the national policy and is likely to further draw attention to the second generation conflict unless immediate steps are taken to address the issues related to institutions and governance regarding community carbon forestry in particular and environmental services provided by forests in general. How forestry capital responsibilities and ownership will change and how to share the income from carbon is bound to take a central stage. This has come at a time when Pakistan is about to draft a new constitu tion along federal lines, which will entail a major restructuring of the administrative boundari es, devolution and decentralization of power. At this juncture, Pakistan's competitive particip ation in the FCPF of the World Bank is to be much welcomed as it provides the financial reso urces enquired to prepare Pakistan for possible participation in REDD in the future.

Biological sequestration of carbon from forest can occur through reduced deforestation or deg radation and better management of forest. In developing countries, including Pakistan, all thes e are primary sources and sinks of carbon. Thus, when forests in developing countries such as Pakistan sequester carbon and implement measures to prevent forest degradation, they need t o be compensated for the emission reduction made possible through one of these means. This was the general argument made by some developing countries in the global climate change di scourse during CoP 6 held in Germany in 2000, especially after realizing the failure of the CD M to address deforestation drivers, leading to loss of large tracts of forest in the tropics (Sikke ma and Kenzie 2001).

Unless a significant change is made in the existing proposal of REDD till Copenhagen in Dec ember 2009, prospect of benefiting Pakistan's forest in general and community forestry in par ticular looks not so encouraging (Dahal and Banskota, 2009). It is mainly due to factors like p oor documentation of deforestation and degradation (uncertainties of baselines), small scale of forests to attract global buyers and Pakistan's low influence in the international REDD policy process, among others. For example, under REDD provision, Pakistan's hilly areas forests where high rates of deforestation and degradation have been registered, could be and appropriate target for Generating Emission Reduction Units (GERUs) by slowing down t he rates of deforestation or degradation by a credible method. However, distribution of benefit s to the hilly areas forest communities alone would b counter productive unless a fair share were provisioned for his co mmunities for restoring and managing their forests in a sustainable way.

Table 1: Prospect of REDD Benefit to Pakistani Forest Sub sectors

Forest Category (Biomass stock and area)	Status of forest in Plan Areas	REDD Prospect* Interest of Intl buyers	REDD Prospect* In country benefit sharing	Status of forest in mountain	REDD Prospect* Interest of Intl buyers	REDD Prospect* In country benefit sharing
Community Managed Forest	Stable/declining	Medium	Medium	Growing/ stable	Low/Medium	High
Government managed Forest	Declining significantly	High	Medium	Stable	Low/Medium	Medium
Protected Areas	Stable or growing	Low	Low	Growing/ stable	Low	Low
Total Forests	Declining	High	Complex & contested**	Stable/ growing	Low/Medium	Contested and risk of perverse***

Stable/ growing

Low/Medium

Contested and risk of perverse***

* Based on original concept of REDD, it is assumed that good performances made against highly deforested and degraded areas will fetch good REDD benefits provided hat other sub   national areas having no negative impacts remained at least a status.

** Lack of clarity of stakeholders and drivers of deforestations, sharing the benefit will be complex as multiple groups may stake their claims.

*** Unless the hill communities received a fair share of REDD benefit for their good performances, there might be risks of perverse impact.

PROBLEMS & CONCLUSSIONS OVER REDD APPLICABILITY IN PAKISTAN

Forest commons are crucial for delivering multiple outcomes such as livelihoods, carbon sequ estration, and biodiversity conservation (Chhatre and Agrawal 2008, GEF 2000, Klooster and Masera 2000, Smith and Scherr 2003). Unfortunately, much of the debate on forest based mitigation measures are dominated by technical issues and socioecological systems suc h as livelihoods and biodiversity tend to take a back seat (Smith and Scherr 2003). Forests pla y a key role in the world carbon cycle (WRI 2000) and also act as a "genetic library" that sup ports important human welfare functions such as the improvement of existing crops, introduct ion of new crops, and the creation of medicines and pharmaceuticals (Myers 1997, Sunderlin et al. 2005). They support the livelihoods of millions of people in the developing world and h elp alleviate poverty while capturing synergy between local and global environment/develop ment goals (Klooster and Masera 2000). For these reasons, the multiplicity of forest commons needs to be an integral part of international debate on forests and climate change.

While CFM has been successful in generating forestry capital, the way of actually claiming ca rbon credits from this is a serious technical issue that still needs to be spelled out by the RED D policy (Karky and Banskota, 2009). There should be three forms of payments for CFM:

a) avoiding deforestation;

b) avoiding degradation and

c) enhancing forest biomass. Whilst the first two are payments for reduced carbon emissions (sources), the third is payment for a carbon sink. For any form of payments to take place, the REDD policy, irrespective of whether it recognizes sink, sources or both, will also have to ad dress

a) how to account for carbon in CFM;

b) what to use as a baseline against which increments are measured; and

c) the rights of indigenous people –

since CFM in most developing countries relates to livelihoods and access to resources (Karky and Banskota, 2009).

The Government of Pakistan has formally entered the readiness mechanism. According to Po kharel and Baral, 2009, there is lacking of the adequate competency of Pakistani professionals for REDD and Capacity of Government in Leading the process of REDD. However, it is not yet known whether community groups and indigenous peoples a re also ready for REDD. A critical mass of community alliance is necessary. More importantl y, community forestry and the protected area system (under which half of the country's forest s are governed) should convincingly be prepared for this concept. Various tenure arrangement s are required that need innovation and piloting of the various models and objectives, ranging from intensive forest management to conservation of biodiversity. The environmental impact of REDD would not be local, but it would be regional or international. It is not clear whether t his would convince local communities and indigenous peoples of the need to participate.

RCOMMENDATIONS

In Pakistan, there is weak mechanism of making up to date database information.
Especially in the case of natural resources, importantly of forest resources there is vast gap. T
he intensive inventory of the whole national forest was carried out before two decades. Havin
g accurate database of the forest resources, carbon status, and deforestation and degradation ra
te is most to enter into the voluntary markets to get benefit from selling carbon credits in the i
nternational markets. Among others, insufficiency of technical manpower is one constraint for
this.
In this scenario the Institute of Forestry like *Shaheed Benazir Bhutto University Sheringal,
PFI, Punjab forestry research institute Faisalabad & Arid zone research institute Bhawalpur*
can take some responsibility to make up to
date inventory of the forest resources and carbon content of Pakistan.
These work can be accomplished through the students of inter to master
&PhD levels with the support from staffs of IOF.
Further
more Institute also can provide some practical knowledge to its students and can evaluate
the work through modifying its syllabi to incorporate more practical classes.
And again MFSC can hire some students from IoF as interns to fulfil its manpower deficiency
in the short as well as long run.

AKNOWLEGDMENTS & DEDICATION

No Achievements is possible without the will & assent of God. His mercy & bounties are countless and whole life is too short for thanking Allah's limitless help and love. The words are too small to thank his numerous mercies.

I am very thankful to all the staff members of the Forestry Department of SHAHEED BENAZIR BHUTTO UNIVERSITY SHERINGAL & PFI PESHAWAR.

I present my special thanks to the Prime Minister RAJA PERVAIZ ASHRAF , Honorable President ASIF ALI ZARDARI, & to the Federal & provincial Ministers of forestry & Environment to be enlighten my aim to *"SERVE NATURE, HUMAN & HUMANITY"*

This Research study/article is dedicated with Deepest gratitude to:

My parents who give me the Freedom to express Myself and taught me the value Of human life & To My mother who enable me to read and go forward after my deceased father and also to My late father Saheeh ul Islam Whose memory will Remain in my heart forever.

REFERENCES

Basnet, R. (2009). Carbon Ownership in Community Managed Forests. *Journal of Forest an d Livelihood, 8 (1)* (Special Issue on Climate change, forestry and local livelihoods), 77 83.

Malik,A,M., & Ahmad,M,R,. (2010) Emporium Forestry MCQ,s and solved papers(pp.201,207)

Chhatre, A., & Agrawal, A. (2008). Forest Commons and Local Enforcement. *PNAS , 105 (36)*, 13286   13291.

Dahal, N., & Banskota, K. (2009). Cultivating REDD in Nepal's Community Forestry: A Dis course for Capitalizing on Potential? *Journal of Forest and Livelihood , 8 (1)* (Special Issue o n Climate change, forestry and local livelihoods), 41   50.

Dhital, N. (2009). Reducing Emissions from Deforestation and Forest Degradation (REDD) i n Nepal: Exploring the Possibilities. *Journal of Forest and Livelihood , 8 (1)* (Special Issue on Climate change, forestry and local livelihoods), 56   61.

DoF. (2009). Community Forestry Programme.Kathmandu, Nepal. *Department of Forest.*

GEF (Global Environment Facility) (2000). Operational Program # 12 Integrated Ecosyste m Management. Available at: http://www.gefweb.org
Gullison, R., Frumhoff, P., Canadell, J., Field, C., Nepstad, D., Hayhoe, K., Avissar, R., Curran, L.M., Friedlingstein, P., Jones, C.D., & Nobre, C. (2007). Tropical Forests and Cl imate Policy. *Science , 316*, 985   986.

IPCC (2007). Working group II Contribution to the Intergovernmental Panel on Climate Cha nge Fourth Assessment Report. Climate Change 2007: Climate Change Impacts, Adaptation a nd Vulnerability. Summary for Policymakers. Available at: http://www.ipcc.ch [Accessed Jul y 10, 2009]

Karky, B. S., & Banskota, K. (2009). Reducing Emissions from Nepal's Community Manag ed Forests:Discussion for CoP 14 in Poznan. *Journal of Forest and Livelihood , 8 (1)* (Special Issue on Cliamte change, forestry and local livelihoods), 33   36.

Karky, B., & Banskota, K. (2007). Case Study of a Community   managed Forest in Lamata r, Nepal. In B. K. K. Banskota (Ed.), *Reducing Carbon Emissions through Community   mana* *ged Forests in the Himalaya.* Kathmandu, Nepal: International Centre for Integrated Mountai n Development.

Klooster, D., & Masera, O. (2000). Community Forest Management in Mexico: Carbon Miti gation and Biodiversity Conservation through Rural Development. *Global Environmental Ch ange , 10,* 259   272.

Kotru, R. (2009). Pakistan's National REDD Framework: How to Start? *Journal of Forest an d Livelihood , 8 (1)* (Special Issue on Climate change, forestry and local livelihoods), 51   55.

Myers, N. (1997). Biodiversity's Genetic Library. In: G.C. Daily (Ed)., *Nature's Services: So cietal Dependence on Natural Ecosystems.* Washington, DC & Covelo, CA: Island Press. [pp. 255   272]

Peskett, L., Brown, D., & Luttrell, C. (2006b). Can Payments for Avoided Deforestation to Tackle Climate Change also Benefit the Poor? *Forestry Briefing 12.*

Peskett, L., Brown, D., & Luttrell, C. (2006a). Making Voluntary Carbon Markets Work Be tter for the Poor: The Case of Forestry Offsets. *Forest Briefing 11 .*

Pokharel, B., & Baral, J. (2009). From Green to REDD, from Aid to Trade: Translating the Forest Carbon Concept into Practice. *Journal of Forest and Livelihood , 8 (1)* (Special Issue o n Cliamte change, forestry and local livelihood), 37   40.

Richard, M., & Jenkins, M. (2007). Potential and Challenges of Payments for Ecosystem Se rvices from Tropical Forests. *Forestry Briefing 16 .*

Rajoka ,A.,Raza,W,H.,Numan,M,A,.(2012)Physical Geography,(pp.187-211)

Sikkema, R., & Kenzie, M. (2001). Market Opportunities for CO2 Credits from Forestry Pro jects. In Carbon Sequestration: Policy, Science and Economics. In E. H. Ryan (Ed.), *Proceedi*

ngs of COFORD Seminar on Carbon Sequestration and irish Forests., (pp. 17 23). Dublin, I
reland.

Smith, J., & Scherr, S.J. (2003). Capturing the Value of Forest Carbon for Local Livelihood
s. *World Development , 31*, 2143 2160.

Staddon, S. (2009). Carbon Financing and Community Forestry: A Review of the Questions,
Challenges and the Case of Nepal. *Journal fo Forest and Livelihood , 8 (1)* (Special Issue on
Climate change, forestry and local livelihood), 25 32.

Sunderlin, W., Angelsen, A., Belcher, B., Burgers, P., Nasi, R., Santoso, L., & Wunder, S
. (2005). Livelihoods, Forests, and Conservation in Developing Countries: An Overview. *Wor
ld Development , 33 (9)*, 1383 1402.

World Bank (2008, July 21). First Countries Named to Benefit from Forest Carbon Partners
hip Facility. *Press release no:2009/029/SDN* . Washington.